MW00736937

THE KID WHO INVENTED THE
POPSICLE

*And Other Surprising
Stories About Inventions*

PUFFIN BOOKS

PUFFIN BOOKS
Published by the Penguin Group
Penguin Putnam Books for Young Readers,
345 Hudson Street, New York, New York 10014, U.S.A.
Penguin Books Ltd, 27 Wrights Lane, London W8 5TZ, England
Penguin Books Australia Ltd, Ringwood, Victoria, Australia
Penguin Books Canada Ltd, 10 Alcorn Avenue, Toronto, Ontario, Canada M4V 3B2
Penguin Books (N.Z.) Ltd, 182-190 Wairau Road, Auckland 10, New Zealand

Penguin Books Ltd, Registered Offices: Harmondsworth, Middlesex, England

First published in the United States of America by Cobblehill Books,
an affiliate of Dutton Children's Books,
a division of Penguin Books USA Inc., 1997
Published by Puffin Books,
a member of Penguin Putnam Books for Young Readers, 1999

3 5 7 9 10 8 6 4 2

THE LIBRARY OF CONGRESS HAS CATALOGED THE COBBLEHILL EDITION AS FOLLOWS:
Wulffson, Don L.
The kid who invented the popsicle : and other surprising stories
about inventions / Don L. Wulffson
p.   cm.
Summary: Brief factual stories about how various familiar things were invented,
many by accident, from animal crackers to the zipper.
ISBN 0-525-65221-3
1. Inventions—History—Juvenile literature. [1. Inventions—History.] I. Title.
T15.W85 1997 609—dc20 96-31148 CIP AC

Puffin Books ISBN 0-14-130204-6

Printed in the United States of America

For my daughters, Jennifer and Gwendolyn, whose questions as little girls were the inspiration for this book.

—DLW

Contents

❑

THE KID WHO INVENTED THE
POPSICLE

Animal Crackers

The design of animal crackers (or cookies) originated in England in the 1890s, but it was in America in 1902 that they really got moving.

In that year, parents across America bought their children a new, inexpensive, and edible toy marketed by the Nabisco Company just before Christmas. The present was 22 animal-shaped cookies in a little rectangular box made to look like a circus cage. The box had a string handle, which made it suitable as a play purse.

That was just a bonus. Initially, the purpose of the string was so that parents could hang the boxes of cookies from the family Christmas tree.

Aspirin

❑

Oddly, the man who invented aspirin did not think it was of much worth or would ever be widely used. French chemist Charles Gerhardt first concocted it in 1853. He did a few quick tests with the medicine, then put it aside. For the next forty years it remained unknown to the world at large.

Finally, in 1893, a young chemist, Felix Hoffman at the Bayer Drug Company in Germany, rediscovered it. Felix's father was suffering from severe arthritis, but none of the known drugs at the time gave him any relief. When Felix happened upon the old French formula for aspirin, he mixed up a batch and gave it to his father. To his astonishment, his father's pain was greatly relieved.

Chemists at Bayer soon began mass-producing aspirin. First marketed in 1899 as a loose powder, aspirin quickly became the world's most prescribed drug. In 1915, the company introduced aspirin in tablet form.

Baby Ruth Candy Bar

The Baby Ruth candy bar was not, as is commonly believed, named after Babe Ruth, the famous baseball player. Originally called Kandy Kake, it was renamed in the 1920s to honor ex-President Grover Cleveland's daughter, Ruth, the first child born in the White House. She was fondly known to the American public as "Baby Ruth."

Interestingly, New York Yankee slugger Babe Ruth *did* try marketing his own brand of candy, Babe Ruth's Home Run Candy. The ball player, however, struck out in the candy business.

Badminton

❏

Badminton originated centuries ago in ancient Babylonia as a fortune-telling ritual. Two people hit a ball back and forth. The length of time the ball could be kept in play supposedly revealed how long the people would live.

In time, the ritual turned into nothing more than a game, and soon was being played in many parts of the world. One place where it was especially popular was at the home of an English duke. Almost every weekend, the duke invited his friends over to play the game.

The duke's home was a mansion—a mansion called Badminton. Little by little, the game took on the name of the place where it was such a popular pastime.

Ball-point Pen

❑

In the 1930s, in Hungary, Ladislao Biro was getting fed up with his old-fashioned fountain pen. He was tired of the way it leaked and had to be endlessly refilled. Fiddling around in his workshop, Biro filled a pen with printer's ink and on the tip of it he fashioned a little ball that picked up more ink as it rolled.

In England, Biro helped set up a factory to manufacture "high-altitude, nonleaking writing sticks" for the British Air Force. The factory was eventually taken over by Bic, a French company, which developed an even better and cheaper throw-away pen.

In America, following World War II, Milton Reynolds invented his own version of the ball-point pen. As a sales gimmick, he advertised it as the "pen that writes underwater." To attract customers, he arranged a demonstration in the display window of a department store. While sitting in a tank of water, a demonstrator scribbled with a ball-point pen on white plastic. In one day, nearly 10,000 had been sold! The price: $12.50 apiece.

Balloon

❑

Toy balloons first came on the market in 1825. Interestingly, they came in the form of a do-it-yourself kit.

The kit consisted of glue, a bottle of liquid rubber, and a little syringe. After putting the glue and liquid rubber in the syringe, the plunger was pushed down and a balloon came out the hollow, needlelike end.

The balloon took three or four minutes to dry. Then it could be separated from the syringe and the rubber stem tied off to keep air from escaping.

Band-Aid

❑

In the 1920s, the Johnson & Johnson Company was in the business of manufacturing large cotton-and-gauze bandages for hospitals and soldiers wounded on the battlefield. One of the employees of the company, a man named Earle Dickson, had a wife who was very accident-prone, frequently cutting or burning herself in the kitchen. Though the injuries were painful and needed tending to, they were far too small to require the company's large sterile dressings. In a moment of inspiration, Dickson cut a little patch of gauze, placed it at the center of an adhesive strip—and invented the Band-Aid.

Johnson & Johnson was soon marketing them, but sales were poor. In a clever advertising gimmick to popularize their new product, the company distributed an unlimited number of free Band-Aids to Boy Scout troops across the country. Sales skyrocketed. The company estimates that since the product was introduced in 1921 more than 120 *billion* Band-Aids have been sold worldwide.

Bar Code Scanner

❑

Bar code scanners, like the type you see in supermarkets, were invented during the 1970s. On packages, as you've noticed, there are black stripes of varying lengths and thicknesses. The scanner basically takes a picture of the stripes. This is relayed to a computer in the cash register, and the computer reads the information as a number.

Types of bar code scanners include gun readers, swipe readers, wand readers, and "glass-top" readers. The average cost of a bar code reader is slightly under $1,000 per unit. The error rate for bar code readers is 1 in 100,000. The human error rate is 1 in 300. Perhaps even more important, the speed of totaling your purchases is up to five times faster.

Barbed Wire

❏

Barbed wire was invented by L. B. Smith of Kent, Ohio, in 1867. The barbs protruded from small blocks of wood strung along the wire strand. The following year a patent was taken out by M. Kelly for a twisted two-strand wire with diamond-shaped barbs. Fences made from "Kelly's Diamond Wire" are still standing in some parts of the United States.

The invention of barbed wire was important to the settling of the "Wild West." In many places there was not enough timber for fencing in cattle ranches. Barbed wire solved the problem.

The other use of barbed wire is on the battlefield. It was first put to this use by U.S. forces in Cuba during the Spanish-American War of 1898.

Barbie Doll

❑

A little hard to believe, but the Barbie Doll started out as a human being! She was Barbara Handler, the daughter of Ruth and Elliot Handler, cofounders of the Mattel Toy Company. The idea for the teenage fashion doll came to Mrs. Handler one day when she noticed that her preteen daughter, Barbie, was losing interest in playing with baby dolls. Instead, she preferred paper cutouts of young women in fashion magazines, even changing their attire by snipping and gluing on changes of clothing.

Barbie Dolls were introduced at the New York Toy Fair in 1959. To date, more than 500 million have been sold.

By the way, the Handlers had a son. His name was Ken.

Baseball Caps

❑

In the 1850s, when baseball was getting its start in this country, most players wore straw hats. This all changed after the Civil War. In the late 1860s, players began wearing visored caps. The caps were imitations of those worn by Union and Confederate soldiers on the fields of battle.

Bikini

❑

On July 1, 1946, the U.S. began peacetime nuclear testing by dropping an atomic bomb on Bikini Atoll. Though it was just a test, it aroused great attention worldwide.

On that day, in Paris, designer Louis Reard was preparing to introduce a daringly skimpy two-piece swimsuit at a fashion show. Capitalizing on media-generated excitement about the atomic test, and believing that his new design was itself explosive, he selected a name then on everybody's lips.

On July 5, four days after the bomb was dropped, Reard's top model, Micheline Bernardini, stepped out into the fashion show lights and paraded down the runway in the world's first bikini.

Bingo

❏

In 1929, toy salesman Edwin Lowe, en route to Jacksonville, Florida, stopped at a carnival. He found one tent full of people seated at tables with numbered cards and piles of beans in front of them. As a man called out numbers, players put beans on the corresponding squares on their cards. The first to get five in a row stood up and shouted, "Beano!"—and won a Kewpie doll.

Upon returning home, Lowe played Beano with his family. One of his children garbled her declaration of winning with "Bean-go!" From then on, Lowe, who helped popularize the game, dubbed it "Bingo."

Blue Jeans

❏

Levi Strauss was a tailor who arrived in San Francisco at the age of seventeen during the Gold Rush of the 1850s. Levi noticed that the miners needed tougher pants, ones that would hold up to the rough work they were doing. Seeing a business opportunity, he stitched tent canvas into overalls. Though coarse and stiff, they held up so well they were soon in great demand.

Not entirely happy with canvas, Levi started using a new fabric, one that was much softer but almost as tough and sturdy. Weavers in the Italian city of Genoa where the fabric was made called it "genes." Strauss changed the spelling to "jeans." To minimize stains, he found that it was best to dye them indigo blue.

Today, "blue jeans" are the best-selling type of pants in the Western world.

Braille

❑

The system of printing and writing for the blind by the use of raised dots felt with the fingers was named "Braille" after its French inventor, Louis Braille. Blinded in an accident in 1812 at the age of three, Louis learned his alphabet by feeling twigs formed in the shape of letters.

When Louis was a teenager, he heard of a French army captain who had invented something he called "night writing"—a system of dots and dashes in relief on thin cardboard by which soldiers could read messages in the dark of night. Inspired, young Louis went to work trying to adapt this system for the use of the blind. After many months of work, he discarded the dashes, and using only dots, devised an alphabet that blind people could read with their fingertips. Not until many years later was Louis Braille's system adopted—first in France, then all around the world.

Breakfast Cereal

❏

One night in the nineteenth century, Ellen Gould White dreamed she met God. He told her that people should lead sinless, healthy lives, and refrain from using tobacco, eating meat, and drinking coffee, tea, and liquor. The dream had such an impact on Miss White that she founded the Health Reform Institute at Battle Creek, Michigan.

Guests were served grain and nut croquettes in place of meat and a cereal-based beverage as a substitute for coffee. The latter was concocted by Charles W. Post, who called it "Postum." He also created a dry breakfast cereal that he called "Elijah's Manna." People liked the taste but not the name, and Post changed it to Grape-Nuts.

Another person at the Institute was Dr. W. K. Kellogg. Primarily to help one of his patients who had broken her false teeth, he made crisp flakes of ground corn. The doctor named them "Corn Flakes," and went on to open the Kellogg cereal company.

Calculator

❑

The calculator was invented by a nineteen-year-old French boy named Blaise Pascal way back in the year 1642. Blaise made it to help his father in his work.

The man was a clerk, and all day long he had to do a tremendous number of mathematical calculations. The boy's invention consisted of a wooden box with sixteen dials on it. By turning the dials, one could do simple addition and subtraction very quickly.

Can Opener

❑

Canned food was invented for the British Navy in 1813. Made of solid iron, the cans usually weighed more than the food they held!

The inventor, Peter Durand, was guilty of an incredible oversight. Though he figured out how to seal food *into* cans, he gave little thought to how to get it out again. Instructions read: "Cut round the top near the outer edge with a chisel and hammer."

Only when thinner steel cans came into use in the 1860s could the can opener be invented. The first, devised by E. Warner of Connecticut, looked like a bent bayonet. Its large curved blade was driven into a can's rim, then forcibly worked around its edge. Stranger yet, this first type of can opener never left the grocery store. A clerk had to open each can before it was taken away!

The modern can opener—with a cutting wheel that rolls around the rim—was invented by William Lyman of the United States in 1870. Sixty-one years later, in 1931, the electric can opener made its debut.

Carousel

❏

The word *carousel* comes from the Italian word *carosello*. A *carosello*, in the seventeenth century, was a tournament consisting of various sports and games. One of the games involved riding on a horse around a circular track and trying to spear a ring. That is why a ride on early carousels—which came to be known as merry-go-rounds in the United States—always involved trying to reach out and grab a brass ring.

Cellophane

❏

Cellophane is not a type of plastic. It is made from a plant fiber, cellulose, which has been shredded, sifted, and aged.

Cellophane was invented in 1908 by Jacques Brandenberger, a Swiss chemist who was trying to make a stainproof tablecloth. But cellophane proved to be far too flimsy to serve as a tablecloth, and ended up as a highly useful food-wrap instead.

Chess

❏

Invented in India in the fifth century A.D., chess was first called *chaturanga*, meaning "four parts." This was in reference to the fact it was a war game in which each player had four types of pieces—elephants, horses, chariots, and infantry—the four basic components of an Indian army at the time.

When the game reached Europe during the Middle Ages, the pieces took on the social and military roles common to European society. Thus, there was the king and queen, knights and bishops, the rook (taken from the Indian word *rukh*), meaning "powerful soldier," and pawns, meaning "foot soldiers."

Chewing Gum

❑

For centuries, the Mayans of Mexico chomped on *chicle*, the dried sap of the sapodilla tree.

In 1845, after he was defeated by the Americans in Texas, Mexican General Santa Anna was exiled to New York. Like many of his countrymen, Santa Anna chewed chicle. One day he introduced it to inventor Charles Adams, who began experimenting with it as a substitute for rubber. Adams tried to make toys, masks, and rain boots out of chicle, but every experiment failed. Sitting in his workshop one day, tired and discouraged, he popped a piece of surplus stock into his mouth. Chewing away, the idea suddenly hit him to add flavoring to the chicle. Shortly, he opened the world's first chewing gum factory.

Gum caught on quickly with Americans. Many doctors, however, said it was unhealthy. In 1869, one wrote that chewing gum would "exhaust the salivary glands and cause the intestines to stick together." Despite such weird warnings, people kept chewing. Today, the average American chews 200 sticks a year.

Cologne

❑

In the Middle Ages, the Black Plague was sweeping across Europe. People were desperate for anything that might save them. Among the many remedies marketed to ward off the plague was a scented, alcohol-based concoction produced in Cologne, Germany, which was to be "drunke and drenched on the body."

In 1709, an Italian barber named Farina emigrated to Cologne. He became interested in the ancient plague remedy, which by then was used by the townspeople as an inexpensive fragrance and after-shave lotion. Putting his barbering aside, Farina went into business producing and marketing *eau de Cologne* ("water of Cologne"), making himself rich and turning the town name into a household word.

Comic Books

❑

The ancient Romans read comics. As early as the first century, tablets with comic inscriptions were sold in the marketplace of Rome.

The first comic book of the kind we know today was published in Connecticut in 1933. Entitled *Funnies on Parade*, it was sixteen pages long and in four colors. It was not sold directly to the public, but was issued as a gift premium by such companies as Proctor & Gamble and Canada Dry.

Coupons

C. W. Post introduced coupons in 1895. In that year, to kick off sales for his new cereal, Post's Grape-Nuts, he distributed little slips of paper offering "one cent off" the purchase price to prospective customers.

Credit Card

❑

The term "credit card," and the concept, came into being before credit cards were in existence. They first made their appearance in 1888 in Edward Bellamy's *Looking Backward*, a novel about a futuristic society in which each citizen was given a yearly "credit card." It was issued by the government and entitled the holder to all necessary goods and services.

In 1900, several U.S. hotels issued "cards of credit" to their best customers. By 1914, department stores and chains of gasoline stations had entered the picture.

Diners Club introduced the first "restaurant card" in 1950. In 1951, the "bank card," today's most popular type of credit card, was introduced by the Franklin National Bank of New York.

Croquet

❑

Croquet was first played in France in the thirteenth century. Known as *pall-mall*, it was not a lawn game. Instead, an iron hoop was set up at the end of a street or alley. Players took turns trying to knock a wooden ball through the ring with a wooden mallet.

In 1850, a group of French people went for a visit to Ireland, and taught the game to their Irish friends. Within two years, the game was popular all over Ireland. Instead of using just one ball, which had been customary, the Irish added more—one ball for each player. They also added more hoops, and also stakes, to be hit as targets. In time, they designed a new mallet, one with a slightly crooked head. They called the game *crokis*, which eventually became *croquet*, a French word meaning "crooked stick."

Crossword Puzzle

❏

The crossword puzzle was invented in 1913 by newspaperman Arthur Wynne. In December of that year, Wynne was working at the *New York World* on its Sunday entertainment supplement called *Fun*. Wynne's editors were pressuring him to come up with a new game feature. Recalling a word puzzle he had played as a child —a simple game called Magic Squares—Wynne was suddenly inspired. Within a few hours he had created the world's first crossword puzzle, which he named "Mental Exercise."

It is not known who came up with the name "Crossword Puzzle." Regardless, within a few years it was a common feature in every major U.S. newspaper. So taken was the public with the game that in the early 1930s women's dresses, shoes, handbags, and jewelry were patterned with crossword motifs.

Regularly enjoyed by more than 60 million Americans today, the crossword puzzle is presently the most popular indoor game in the country.

Dice

❏

The origins of dice reach back over 40,000 years! Their earliest use was not for games and gambling, but for fortune-telling. The dice, which came in a variety of shapes and which had no dots, were shaken in the hand and thrown to the ground. The way the dice lay, in relation to each other, supposedly told how long a person would live, how happy he would be, and so forth.

Dice are among the oldest fossil remains of inventions used by early man. Archaeologists have found dice shaped from the ankle bones of sheep and dogs. The early inhabitants of North America crafted dice from peach pits, buffalo bones, deer horns, stones, walnut shells, and beaver and woodchuck teeth.

The Egyptians added dots to dice and were the first to use them for gambling. As for the origins of cheating with "loaded dice," it is as old as the objects themselves. Many of the first Egyptian gaming dice discovered by archaeologists were "loaded"—that is, weighted to favor one side over the others.

Dominoes

❑

A game similar to dominoes was played in ancient China using playing cards with dots on them. The dots represented all the possible throws with two dice.

There is no record of the game being played in Europe until the Napoleonic Wars of the eighteenth century. During these times, French prisoners of war imprisoned in England played a game using homemade black-on-white tiles. The tiles reminded the players of a garment worn by priests, a dark, hooded robe lined in white. The robe was called a *domino*. The prisoners soon began referring to the game they were playing in its plural form: *dominoes*.

Doughnut

❑

The doughnut was "invented" in Holland in the sixteenth century, but strangely, over 250 years passed before the hole appeared in the middle.

About the size of a walnut, Dutch bakers called them *olykoek,* or "oil cake," due to the fact that they were deep-fried in oil. When the Pilgrims lived in Holland, they learned to make "oil cakes," which they called "dough nuts." When they left Holland and came to America in 1620, the Pilgrims brought the recipe with them.

One day in 1847, Hanson Gregory, a sea captain who had just returned to his home in Maine from a long sea voyage, found his mother in the kitchen getting ready to fry a batch of doughnuts. The captain was very fond of the pastry but, like many people, he was bothered by the fact that the center was always soggy. Before his mother fried the doughnuts, he took a knife and cut holes in them. When they were done, he found that the soggy center had been eliminated—and the doughnut had taken on a whole new shape.

Dr. Pepper

❑

Hires Root Beer was named after the man who concocted it in 1876, Charles E. Hires. Dr. Pepper has a stranger story behind how it was invented and named.

Working at a pharmacy in Virginia around 1880 was a young man (whose name has been lost to posterity). The young man, a "soda jerk," fell in love with the pharmacist's daughter. Putting a quick end to the romance, the pharmacist fired the young man and told him to get out of town and never show his face again.

Jobless and brokenhearted, the young man went west, eventually ending up in Waco, Texas. There he got a job, naturally enough, at a local soda fountain. In time, the young man became well-known for two things: the story of his unhappy romance back in Virginia and his knack for brewing up original soft drinks.

As a joke, the young man's customers named his best soft drink after the pharmacist who had caused his broken heart. The pharmacist's name was Dr. Pepper.

Drinking Straw

❑

The Greeks and Romans used drinking straws made of hollow glass tubes. Drinking straws of waxed paper appeared in the United States and Britain in the nineteenth century, to be replaced in the mid-twentieth century by plastic straws.

Electric Blanket

❑

In 1912, American inventor S. I. Russell patented an electric heating pad. Its sole purpose was to warm the chests of patients suffering from tuberculosis. The pads were relatively small, and the wiring was so inadequate that they were liable to burst into flames.

For many years, inventors tried to come up with a full-size electric blanket. But it was not until World War II that they got anywhere. During the war, the air force invented electrically heated flying suits for pilots. Using the same technology and principles, American inventors soon turned the electric blanket into an everyday reality.

Elevator

❑

The first elevator was built in 1743 for King Louis XV at his palace in France. The one-person contraption went up only one floor, from the first to the second. Known as the "Flying Chair," it was on the outside of the building, and was entered by the king via his balcony.

The mechanism consisted of a carefully balanced arrangement of weights and pulleys hanging inside a chimney. Men stationed inside the chimney then raised or lowered the Flying Chair at the king's command.

Fig Newtons

❑

In 1895, a new machine was installed at a Massachu-setts cookie company that could wrap cookie dough around jam. The first jam the operators tried with the machine happened to be "preserve of figs." Company policy was to name new products after neighboring towns. Since the little town of Newton was nearby, the name Fig Newton was adopted.

Flashlight

❑

The flashlight began as a novelty item called the "electric flowerpot." It consisted of a slender battery in a tube with a light bulb at one end. The tube rose up through the center of a flowerpot and illuminated an artificial plant.

When the novelty item bombed, the inventor found himself with a huge overstock. Attempting to salvage a little of his investment, he separated the light and tube (which was made of cardboard) from the pot. He was soon selling what he dubbed the "Portable Electric Light"—and made a fortune.

Flyswatter

❏

In 1905, schoolteacher Frank H. Rose of Topeka, Kansas, made a fly-smashing device by attaching a square of wire screen to the end of a yardstick. Initially, Rose called his invention the "fly bat," but changed it to "flyswatter" at the suggestion of a friend. The holes in the wire screen were essential because a fly can sense the air pressure of a solid object, such as a rolled-up newspaper, coming at it.

Foam Rubber

❏

During the 1920s, inventor E. A. Murphy of England beat latex rubber into a foam, using an egg whisk. Later, Murphy and fellow inventor W. H. Chapman used a food mixer, then poured the gel-like foam into molds.

The first foam rubber products were motorcycle seats, produced in 1931. The following year, foam rubber was used to make seats for 300 London buses and to upholster the seats for the new Shakespeare Memorial Theatre at Stratford-on-Avon.

Frisbee

❑

The Frisbee is named after the Frisbie Pie Company, a Connecticut bakery that first opened for business during the 1870s. For fun, people tossed around the empty pie tins, which had the name "Frisbie" embossed on the bottom.

In the early 1950s, a man by the name of Walter Morrison devised a metal tossing toy he called the "Flying Saucer." Soon, because metal could be dangerous, he changed the metal to plastic. Then he changed the name. Morrison, at the time, was living in California, but recalling his younger days in Connecticut where he and his friends had tossed around pie tins from the local bakery, he decided to name his invention the "Frisbee."

Gatorade

❏

In the 1960s, Dr. R. Cade was making a study of heat exhaustion and its effects. His test group was the football players at the University of Florida, whose team name is the "Gators." After only three minutes of analyzing the body liquids lost as a result of sweating, Dr. Cade came up with the formula for Gatorade.

After two years of trying to sell the product on his own, Dr. Cade sold the formula for a meager sum to the Stokely-Van Camp Company. Soon, sales were exceeding $50 million a year, and Gatorade could be found on the training tables of thousands of high school, college, and pro sports teams across America.

Hair Dryer

❑

The first hair dryer was the vacuum cleaner! Around the turn of the century, women dried their hair by connecting the hose to the exhaust of their vacuum cleaners. In early models, the front of a vacuum cleaner sucked air in, the back blew air out, and the hose could be attached to either end.

In 1920, the first true hair dryer came on the market, but it was extremely large and heavy, and frequently overheated. Not until 1951 was the first really workable dryer made. The device consisted of a hand-held dryer connected to a pink plastic bonnet fitted over the woman's head.

Hearing Aid

❑

The largest hearing aid of all time was the "acoustic throne," built for King John of Portugal in 1819. Kneeling before the throne, people spoke into the hollow arms of the chair, and resonators passed the sound through tubes inserted in the royal ears.

During the nineteenth century, there were other strange "sound-amplifying devices." For hard-of-hearing men, there was the "acoustic walking stick," the "acoustic top hat," and an under-the-beard receptor, all with tubes to the ears. For women, there were hearing aids disguised as bonnets, crowns, and handbags.

The first battery-powered hearing aid, the Otophone, was marketed in 1923 by the Marconi Company. To say the least, the Otophone was cumbersome. The equipment was packed in a case weighing 16 pounds! By 1930, the device had been reduced to 4 pounds; by 1935 it was down to $2^{1}/_{2}$ pounds. By the 1950s, the invention of the transistor made possible the development of hearing aids that were almost weightless and nearly invisible.

High Heels

❏

A man, not a woman, was the first to wear high heels. King Louis XIV of France was a very short man. Tired of having to look up to other people, he ordered the royal cobbler to make him a pair of shoes with extra-high heels, and soon, both men and women were wearing them. The fashion gradually faded with men, but not with women.

Hot Dog

❑

In the early 1900s, the streamlined sausages with the almost transparent casing were becoming popular in the United States. They went by a variety of names: frankfurters, franks, wieners, red hots, and dachshund sausages.

One day in 1906, cartoonist Tad Dorgan was in the stands at a New York Giants baseball game. Inspired by the vendors' call of "Get your hot dachshund sausages!" Dorgan went back to his office and sketched a cartoon of a dachshund dog in a bun and smeared with mustard. Not sure how to spell "dachshund," Dorgan left the word out when he captioned his cartoon: "A real hot dog!"

The cartoon, appearing nationwide, added a new, all-American word to our vocabulary.

Hot-air Balloon

❑

The hot-air balloon was invented by Jacques and Joseph Montgolfier, two French brothers who ran a paper bag factory. They came up with the idea one day when they filled a paper bag with hot air and found that it floated. In 1783, the brothers went for a test flight. When they landed in a field, terrified peasants thought they were under attack by creatures from another planet, and proceeded to hack the balloon to pieces.

Hula Hoops

❏

More than three thousand years ago, children in Egypt played with large hoops of dried grapevines. The toy was propelled along the ground with a stick or swung around at the waist.

During the fourteenth century, a "hooping" craze swept England, and was as popular among adults as kids. The records of doctors at the time attribute numerous dislocated backs and heart attacks to "hooping." The word *hula* became associated with the toy in the early 1800s when British sailors visited the Hawaiian Islands and noted the similarity between "hooping" and hula dancing.

In 1957, an Australian company began making wood rings for sale in retail stores. The item attracted the attention of Wham-O, a fledgling California toy manufacturer. They changed the wood rings to plastic, and in January of 1958 introduced Hula Hoops to the American public. Within six months, Americans bought 20 million hula hoops, at $1.98 apiece.

Ice-cream Cone

❑

It happened at the 1904 World's Fair in St. Louis, Missouri. One especially hot afternoon, a teenage ice-cream vendor, Arnold Fornachou, had so many customers that he ran out of paper ice-cream dishes. Desperate, he went to other nearby vendors to see if any might have some spare containers. None did. But then Arnold happened to pass by a stand where an Arab man was selling *zalabia*, a wafer-thin, wafflelike confection sprinkled with sugar. The teenager bought a stack of them, and soon was selling scoops of ice cream plunked into the tops of *zalabia* twisted into the shape of a cone.

Cones were rolled by hand until 1912, when F. Bruckman, of Portland, Oregon, invented a machine for doing the job. Today, one-third of all ice cream eaten in the United States comes in a cone.

Jigsaw Puzzle

❏

"For the purpose of teaching geography," John Spilsbury, a teacher in England, created the first jigsaw puzzle in the year 1767. Hand-painted and made of wood, the puzzle was a map of England and Wales, with each county making up a separate piece.

Soon, people began making pictorial jigsaw puzzles. Their purpose was to entertain rather than to teach.

The pieces in these early puzzles were not interlocking. Not until the invention of power tools more than a century later did jigsaw puzzles with fully interlocking pieces come into being.

Kleenex

❏

The first use of what we know today as Kleenex was as an air filter in GIs' gas masks during World War I. After the war, the company making the substance had a huge surplus in its warehouse. Needing a market for this surplus, company executives came up with a product they called Celluwipes, which they advertised and sold as a "Sanitary Cold Cream Remover." The name Celluwipes was soon changed to Kleenex.

Woman liked the product. Many wrote letters to the company praising it as a wonderful cold cream remover. But they also complained that their husbands and kids were blowing their noses on their cold cream wipers. The company got the message. Kleenex was remarketed, plugged primarily as a tossable hanky.

Lawn Mower

❑

In 1830, Edwin Budding of England patented a "machine for shearing the vegetable surface of lawns, grass plots, and pleasure grounds." Called a Roller Mower, it was rather clumsy and heavy, but looked and functioned much like today's machine, except that the collection box resembled a seed tray.

In 1919, an American Army colonel "borrowed" the motor from his wife's washing machine, and produced the world's first gas-powered lawn mower.

Marshmallow

❏

The mallow was one of the most common plants in Medieval England, and one variety was so hardy that it even grew in salt marshes. From the roots of this "marsh mallow," syrup was extracted that was used in making medicines. According to a medical book published in 1680, it was the remedy for half a hundred illnesses.

Early in the eighteenth century, it was discovered that mixing gum arabic with marsh mallow syrup created a soft, puffy, candylike confection. Soon, a substitute was developed from gelatin and sugar. Though the confection today has absolutely nothing to do with the plant, it is still known as a "marshmallow."

Matches

❏

One day in 1826, a man named John Walker was in his workshop in England trying to invent a new type of explosive. After stirring a mixture of chemicals with a wooden stick, he put the stick down and went to tend to other business. When he returned, he found that a hard little glob had dried on the tip of the stick. Trying to remove it, he scraped it against the stone floor of his shop. To his surprise, the stick ignited—and the friction match was born.

Mickey Mouse

❑

Believe it or not, Mickey Mouse started his career as a rabbit—in 1927 as *Oswald the Lucky Rabbit*, to be exact. Oswald was a black rabbit with a white face. Except for the wrong kind of tail and ears, he looked a lot like Mickey Mouse.

Because of legal difficulties as to who had the rights to Oswald, Walt Disney decided to change the name, and eventually the type of animal. In his studio in Kansas City, Disney had a tame mouse that he had nicknamed Mortimer. Soon, Disney was drawing Mortimer Mouse cartoons. However, before any of these were released, the name was changed to Mickey Mouse.

Minnie Mouse, with no aliases, was with Mickey almost from the start. Pluto, Mickey's dog, did not make his appearance until 1930, and was originally called Rover.

Microwave Oven

❏

The microwave oven did not come about as a result of someone trying to find a better, faster way to cook. During World War II, two scientists invented the magnetron, a tube that produces microwaves. Installing magnetrons in Britain's radar system, the microwaves were able to spot Nazi warplanes on their way to bomb the British Isles.

By accident, several years later, it was discovered that microwaves also cook food. Called the Radar Range, the first microwave oven to go on the market was roughly as large and heavy as a refrigerator.

Miniature Golf

❑

John Carter of Chattanooga, Tennessee, loved golf, but he wasn't very good at it. He also loved spending time with his family, but it seemed there often wasn't anything fun that a family could do together.

One day in the late 1920s, he suddenly had an idea. How about a little golf course designed just for families to have fun? Excited, Carter set about building what he called the "Tom Thumb Golf Course." Players could only use putters. And there were all sorts of silly obstacles and fun ways of trying to get the ball into the cup.

Carter's "Tom Thumb" course was a real hit. Soon, imitators across the country were opening up their own versions of what came to be called "miniature golf."

Monopoly

❑

One day during the Great Depression, Charles Darrow made a buying-and-selling game. Hand-painting it on a piece of linoleum, he wrote the names of actual streets in Atlantic City, New Jersey, on the board—Marvin Gardens, Ventnor Avenue, Boardwalk. At first, he only played "Monopoly" at home with his family and friends. But others soon heard of the game and ordered sets of their own. Darrow went to work, making handmade copies of Monopoly at $4.00 apiece.

Darrow took his invention to the Parker Brothers Company. At first, he was turned down (by brothers Charles and George Parker), but they soon reconsidered. The company sold every Monopoly set they could manufacture for Christmas, 1934, and the demand did not die down after the holidays. So many orders came in that they were stacked in baskets in hallways.

Within a short time, Monopoly was the best-selling game in the United States—and Darrow was a millionaire many times over.

Motion Pictures

❑

Though Thomas Edison took credit for inventing movies, the credit rightfully belongs to a man named William Dickson.

In 1888, Edison was seeking a way to add pictures to his recent invention, the phonograph. One of his employees, William Dickson, figured out how to do it. Called a Kinetoscope, the device consisted of a wooden box in which a single viewer could watch a "peep show."

Edison *did* open the first movie theater—on April 14, 1894. But the "theater" was nothing but a bunch of Kinetoscopes, each for a solitary viewer. Edison personally opposed projecting pictures on a large screen, believing it had no future!

The first inventors to project a picture on a screen were two French brothers, Louis and Auguste Lumière. On December 28, 1895, they showed a wide-eyed crowd the first real movie. In it, a train enters a station, a boat is rowed out of a harbor, and workers walk out of a factory at quitting time.

Motor Scooter

❑

The first motor scooter, the Auto-Ped, was put on the market in 1915 in New York. It resembled a child's toy. There was no seat, and the rider had to stand on a low platform while driving the machine.

Motor scooters were used during World War II by paratroopers. The scooters were dropped with parachutes, so that soldiers could get away from the enemy quickly. The scooters had machine guns mounted in front. They looked very dangerous—and they were.

Motorcycle

The first motorcycle, built in 1884 by an Englishman named Edward Butler, looked pretty silly. It had three wheels, not two, and was really just a tricycle with a motor.

Nevertheless, people were afraid of Butler's motorcycle—so afraid that they asked the government to pass laws against the new machine. One law said that there must always be three people on a motorcycle. Another said that a man with a red flag must run ahead of the motorcycle, waving the flag and yelling to warn people that a motorcycle was coming.

At about the same time, a German named Gottlieb Daimler invented another kind of motorcycle. Daimler's motorcycle had only two wheels—but the wheels, and the rest of the contraption, were made entirely of wood! Paul Daimler, Gottlieb's young son, was the first to give his dad's motorcycle a test drive. His daughter is also said to have taken it for a spin, but cracked it up into a tree.

Neon Lighting

❑

In H. G. Wells' sci-fi novel, *When the Sleeper Awakes*, published in 1899, many of the scenes are lit by neon lights. The strange thing is, neon lights hadn't been invented yet!

Eleven years after publication of the book, in 1910, Georges Claude, a French scientist, made neon lighting a reality. Claude's problem with his neon tubes was that the light they gave out was always red. After years of experimenting, it was found that the color could be changed by mixing powders of different hues into the glass tube.

Paper Bag

❑

There are few things simpler and more functional than the paper bag. Americans use them at the rate of 40 billion a year.

In 1883, after fighting for the Union in the Civil War, Charles Stilwell began to tinker with the idea of making a better paper bag. Paper bags already existed at this time, but they had many flaws. They had to be pasted together by hand; their V-shaped bottoms prevented them from standing on their own; and they were not easily collapsible or conveniently stackable.

In the summer of 1883, Stilwell put into operation the first machine to produce paper bags—bags far superior to anything that existed. The bags had flat bottoms for standing up straight by themselves and pleated sides that made them easy to fold and stack. With the birth of the American supermarket in the early 1930s, demand for Stilwell's paper bags skyrocketed. Their versatility, strength, and low cost made them a nationwide—then worldwide—phenomenon.

Paper Cup

❑

The story of the paper cup begins in 1908. A young inventor, Hugh Moore, produced a vending machine to dispense a cup of pure, chilled drinking water at a penny a serving.

To Moore's disappointment, nobody was interested in paying for a cup of water. At that time, people drank water out of a bucket, using a common tin dipper. Because the dipper was used by the sick and the healthy alike, and was seldom washed or sterilized, it was a real health hazard. By chance, Moore met a rich man who greatly disliked the unsanitary dippers, and who, after one meeting, decided to invest $200,000, not in Moore's water-vending machines, but just in his paper cups. Overnight, Health Kups were invented.

Moore's office happened to be in the same building as the Dixie Doll Company. He liked the name, and in 1919 changed Health Kups to Dixie Cups.

Paper Towels

❏

In 1907, the Scott brothers' paper company was a big success in making and selling toilet paper. Their high-quality tissue arrived from a paper mill in the form of what were called "parent rolls"—huge rolls of toilet tissue which were then cut down to convenient bathroom-size packages.

One day an order arrived from the mill that was clearly defective. The parent roll was excessively heavy and wrinkled. It was scheduled to be returned when a member of the Scott family came up with the idea of perforating the thick paper into towel-sized pieces. A few weeks later, Sani-Towels, the world's first disposable paper towels, went on the market.

Parachute

❑

During the eighteenth century, many inventors tested parachutes, and some died as a result of their failures. One of the most bizarre was the "parachute hat," a helmet to which a parachute had been attached. For a soft, bouncy landing, the inventor wore big rubber bouncy shoes. Unfortunately, in his one and only test of the "parachute hat," the inventor broke his neck.

Jacques Garnerin made the first successful flight to earth in 1797 in France. He ascended above Paris in a basket suspended from a hot-air balloon. Also attached to the basket was an umbrella-like parachute. Reaching a height of 3,000 feet, Garnerin cut the cords attaching the balloon to the basket, then descended safely to the ground as the basket swayed gently back and forth, supported by his parachute.

Peanut Butter

❏

Late in the 1880s, a St. Louis doctor was facing a complicated problem. One of his patients was gravely ill from "protein malnutrition," but a severe stomach disorder prevented the man from eating meat.

Because they are high in protein, the doctor ground roasted peanuts into a paste. When peanut oil was added to the dry mixture, it proved to be a palatable, readily digestible, and lifesaving remedy for his patient.

As a medicinal remedy, the concoction generated so little interest that the name of the doctor and his patient were not recorded. But soon, news of the tasty new food—"peanut butter"—was spreading from coast to coast, and quickly became a staple on grocery shelves everywhere.

Pencil

❏

The first pencils consisted of chunks of graphite wrapped in string. As the graphite was used up, more of the string was unwound. Next came tubes of leather or wood, into which pieces of graphite were pushed, shortly followed by metal holders with clawlike ends for holding the graphite in place.

Wooden pencils came into being in 1683. In that year a man named J. Pettus took a small stick of cedar, slit it down the middle, hollowed out the center, inserted a piece of graphite, then glued the whole thing together to create a wooden pencil.

For a long time, all wooden pencils were square. Rounded pencils were invented in 1876 by an American named Joseph Dixon.

The first factory for making pencils was built in 1662 in Germany by a man named Kasper Faber. In 1848, Eberhard Faber, a great-grandson, emigrated from Germany to the United States where, in 1861, he built the first pencil factory in this country.

Phonograph

❏

Though conceived and designed by Thomas Edison, the first phonograph was built by his mechanic, a man named John Kruesi. Edison's first message, "Mary had a little lamb," was recorded on a tinfoil cylinder on November 20, 1877.

Soon Edison had vastly improved the phonograph. The earliest records were on wax cylinders. The first musical recording was made by Jules Levy, playing "Yankee Doodle" on a cornet. The first instrument for domestic use and sale was the Edison Parlor Speaking Phonograph, marketed at $10 in 1878.

Edison was so excited about his new invention that he came up with a rather strange idea. He wanted to put a large phonograph in the mouth of the Statue of Liberty to give it a voice that could be heard all over New York Harbor. His idea was met with silence.

Ping-Pong

❏

When table tennis was invented in England during the nineteenth century, all the equipment was homemade. The ball consisted of wound-up string. Books put down the middle of a table made the first net. The paddle was cut out of a piece of thick cardboard.

Balls of rubber or cork soon replaced those made of string. One day a man named James Gribb noticed children playing with hollow celluloid balls. He tried them out at table tennis and found they were perfectly suited to the game.

The surface of early wooden Ping-Pong paddles was too slick and didn't give the player enough control. A man named E. C. Goode noticed a rubber cash mat on the counter in a drugstore. Suddenly the idea came to him that it would make a good surface for a Ping-Pong paddle. Goode bought the mat, hurried home, cut it down to size, and glued it to the paddle. The new paddle improved his game so much that he challenged the British national champion—and won!

Pockets

❑

Pockets were not invented until the seventeenth century. Until that time, a man carried his belongings in a pouch hung from a loop in his belt. Interestingly, the first pockets for pants were like pouches; they were made to hang *outside* the man's pants, not inside! Not until the eighteenth century did pockets inside of pants come into being.

Popsicle

❏

One day in the winter of 1905, eleven-year-old Frank Epperson mixed a jar of powdered soda pop mix and water. Accidentally, he left the mix on his back porch that night. The next morning Frank found the stuff frozen, with the stirring stick standing straight up. Pulling out the frozen soda pop, stick and all, he realized he had accidentally invented something pretty good.

Calling it the "Epperson Icicle" (which he soon changed to "Epsicle"), the next summer he made them in the family icebox and sold them around the neighborhood at five cents apiece. Later, he renamed his product the "Popsicle," since he'd made it with soda pop.

Postage Stamps

❏

Until the eighteenth century, there was no organized postal system. Some letters had crude stamps on them, others bore handwritten symbols, and others had no postal markings at all. The fee for handling letters was paid by the person *receiving* a piece of mail, not by the sender. This all changed in 1840 with the invention of the adhesive postage stamp by Englishman Roland Hill. From that time on, postage—in the form of stamps—was paid by the sender.

Potato Chips

❑

In the summer of 1853, American Indian George Crum was a chef at a fancy restaurant in New York. One guest ordered French fries, which at the time were flat wedges of fried potatoes. The man kept sending back his fries, complaining that they were too thick. Annoyed, Crum decided to play a trick on the guest by producing French fries that were ridiculously thin and crisp. Instead of angry with Crum's little practical joke, the guest was ecstatic over the browned, paper-thin fries. Soon "potato chips" had become a specialty of the restaurant.

In the 1920s, a salesman named Herman Lay began traveling the country selling bags of potato chips out of the trunk of his car. Within a few years, Lay was rich and Americans everywhere were eating the crisp, salty snack that had originated as a practical joke.

Pretzel

❑

In the eighth century, the Saxons, a Germanic people, were conquered by Charlemagne, the first emperor of the Holy Roman Empire. Charlemagne was a Christian, and he forced the Saxons to convert to Christianity. Not only did he demand that they go to church and celebrate Christian holidays, he also ordered them to bake the sign of the cross into their bread.

At first, this consisted of two crossing strips of dough on top of the loaf. In time, the strips of dough were baked separately, with the cross enclosed in a circle. The pretzel was the result.

The word itself comes from the Latin term *pretiola*, meaning "a little reward," since it was customary to give children the treat as a reward for saying their prayers.

Quiz

❑

The word *quiz* came into being in a very unusual way. One day in the late eighteenth century, an Irishman named Daly bet a friend that he could introduce a new word into the language in 24 hours. That night Daly went all around the city of Dublin writing QUIZ on walls. The next day everyone in Dublin was asking, "What is a quiz?" Daly won his bet, and the word *quiz* became a part of the English language—literally overnight. Because Daly's quiz tested people's ability to come up with a quick answer, the word has come to mean a short, fast test.

Rear-view Mirror

□

There were forty cars entered in the Indy 500 race in 1911. It was customary at that time for two men to be in each car. One was the driver. The other was the mechanic, who also served as a "lookout." He was responsible for seeing what was going on behind the car.

When Roy Harroun showed up for the race, he let it be known he was driving alone. The other drivers protested, and said it would be unsafe to drive alone without a lookout.

Harroun solved the problem in an interesting way, and in a way that would change cars forever. He devised a special mirror, which he attached to his car. He called it a "rear-view" mirror.

The judges agreed that the mirror served the same function as having a lookout, and Harroun was allowed to compete. And all by himself, six hours and 42 minutes after the race began, he roared across the finish line in first place.

Reflecting Road Studs

❏

A cat was the inspiration for those little reflectors you see on streets and roads.

One night in 1933, Englishman Percy Shaw was driving home. It was extremely dark and foggy. Suddenly, ahead, Shaw's headlights caught the reflection of a cat's eyes. Stopping his car, he realized that he had been driving directly at a fence on which a cat was sitting, and on the other side of the fence was a sheer drop of hundreds of feet!

After a year of experiments, Shaw installed the first fifty reflecting studs at his own expense at a spot notorious for the many accidents that had occurred there. Soon, what he called "Follsain Gloworm Studs" were being installed in England and in many countries around the world.

Refrigerator

❑

Believe it or not, the ancient Greeks and Romans had crude refrigerators. They transported snow from mountaintops into their homes, and put it in a "snow cellar." This was a hole dug in the ground, lined with logs, insulated with thick layers of straw, and packed with snow. The compressed snow turned into a solid block of ice, which remained frozen for months, providing primitive but effective refrigeration for perishable foods.

The first modern refrigerator for domestic use was the Domelre, manufactured in Chicago in 1913. The Domelre consisted of a heavy wooden cabinet with an electrically powered refrigeration unit mounted on top.

Remote-control Toys

❑

During World War II, the Germans developed remote-control boats, planes, and miniature tanks. These were packed with bombs, and remote controls guided them toward the enemy. When they reached their target, the bombs were exploded. It was from these weapons of war that the idea for remote-control toys came into being during the early 1950s.

Robot

❏

The word "robot" was coined by Czech playwright Karel Capek in his 1920 play *R.U.R.* In the play, Capek showed lifelike robot workers taking over the world.

Forty years later, in the early 1960s, robots by the millions took the world by storm—mostly in the form of toys. However, today we are beginning to see the use of robots in police work, in science labs, on factory assembly lines, and for space exploration.

Roller Skates

❏

The first pair of roller skates, called "wheeled feet," was the brainstorm of a Belgian instrument maker, Joseph Merlin, in 1759. Each skate had only two wheels, which were made of wood and aligned along the center of the shoe.

Merlin constructed the skates to make a spectacular entrance at a costume party in the Belgian city of Huy. A master violinist, he intended to roll into the party while playing his violin. Unfortunately, he hadn't yet mastered the art of stopping. Skates strapped to his feet, playing away on his violin, Merlin came rolling into the party— and crashed into a full-length mirror, breaking it, his violin, and his "wheeled feet."

Root Beer

❏

American colonists became acquainted with the sarsaparilla plant, the roots of which, when boiled, yielded a homemade cough syrup.

In 1866, a medical student in Philadelphia became interested in the syrup. He concluded that it had no medicinal value, but he was intrigued by the novel flavor. After months of experimentation, he succeeded in preparing a tasty beverage flavored with sarsaparilla extract. Because it came from the roots of the plant and looked a bit like beer, he called it "root beer."

The young man's name was Charles Elmer Hires. His root beer was first sold in a Philadelphia drugstore at five cents per mug. It made such a hit that Hires put other interests aside, and in 1876 started a national business marketing and selling Hires Root Beer.

Rubber Gloves

❑

During the nineteenth century, a surgical nurse at Johns Hopkins Hospital was suffering from rashes on her hands as a result of daily preoperative scrubbings in an antiseptic solution. Dr. William Halstead, chief of the hospital's surgical staff, became aware of the nurse's problem and took it upon himself to find a solution.

He made plaster casts of her hands, then took the casts to a manufacturer of rubber products and had thin gloves molded from them. The gloves not only solved the immediate problem, they also fostered a romance between the two which eventually led to marriage. In time, Dr. Halstead tried the gloves himself, found they were well-suited to surgery, and ordered a pair of his own. More sterile than even the best-scrubbed hands, rubber gloves were quickly adopted by surgeons everywhere.

Safety Glass

❏

Safety glass was invented by accident. One day in 1903, French chemist Edouard Benedictus accidentally knocked a glass flask to the floor. To Benedictus' surprise, the broken pieces of the flask still clung together.

From his assistant, Benedictus learned that the flask had contained a solution of liquid plastic. This had evaporated, leaving a thin, transparent coating inside the flask.

Excited about his discovery, the chemist tried to interest automakers in what he called Triplex, explaining how it could be used as windshields to save motorists from needless injuries. Incredibly, they were not interested! They did not want to incur the small added expense, and basically said that it was not their problem if drivers and their passengers got hurt in accidents.

During World War I, the U.S. Army began using safety glass for lenses in gas masks. Finally, auto manufacturers realized their mistake and began making windshields out of the material—as Benedictus had suggested they do more than a decade before.

Safety Razor

❑

At the age of twenty-one, King C. Gillette was a traveling salesman with a yearning to be an inventor. A friend, William Painter, inventor of the bottle cap, advised Gillette to invent something that people use a couple of times and then throw away.

Gillette became obsessed with inventing a disposable item. The quest lasted years, but it finally paid off. Gillette described the historic moment: "As I stood there with the razor in my hand, my eyes resting on it as lightly as a bird settling down on its nest, the Gillette Safety Razor was born—more with the rapidity of a dream than by a process of reasoning. In that moment I saw it all: the way the blade could be held in a holder; the idea of sharpening the two opposite edges on a thin piece of steel; the clamping plates for the blade, with a handle halfway between the two edges of the blade."

Sandwich

❏

During the eighteenth century, John Montagu, England's fourth Earl of Sandwich, was so addicted to gambling that he refused to leave the gaming tables, even for meals. So that he could keep playing while he ate, he ordered his cook to serve him sliced meats and cheeses between two pieces of bread. His well-known habit of snacking in this way resulted in the sandwich being named for him.

Scooter

❑

The scooter was invented in 1897 by Walter Lines, a fifteen-year-old English schoolboy. Since the boy's father did not think the contraption was worthwhile, a patent was never taken out on it. When Walter grew up, he opened a toy factory, Triang Toys. The scooter was his best-selling item.

Scotch Tape

❑

In 1925, some Detroit carmakers were putting out two-tone models, which were a nightmare to paint. The carmakers turned to the Minnesota Mining and Manufacturing Company (3M) for a sturdy tape that could be used to keep the borders clean and straight where the colors met. The people at 3M invented cellophane tape. But they skimped—sending off batches of tape with adhesive only on the edges, not down the middle. It stuck poorly, and paint bled through.

Disgusted, the autoworkers complained to 3M salesmen for sending them "Scotch" tape, the word "Scotch" at the time being a slang term for "cheap." Said the workers: "Take this Scotch tape back to those Scotch bosses of yours and tell them to put adhesive everywhere on it." The bosses gave in to the workers and made the tape right, but the name "Scotch" stuck.

Scrabble

❑

In 1931, Alfred Butts translated his lifelong love of crossword puzzles into a board game. The game, which he called Criss-Cross, was strictly a home entertainment for his family and friends. Butts had never even considered selling the game until James Brunot, a good friend of Butts, made the suggestion in 1948. Brunot even thought of a new name for the game: Scrabble. The name, he thought, was sort of like the game itself. The players scrabbled, or scraped together, tiles, then struggled to make the scrambled letters into words.

Eventually, the two men went into business manufacturing the game in an old schoolhouse. For four years they eked out a living. Then in the summer of 1952, for no readily apparent reason, Scrabble suddenly became a fad. Sales skyrocketed from less than 10,000 games a year to more than four million a year. Today, Scrabble ranks (just below Monopoly) as the second best-selling game in U.S. history.

Shopping Cart

❑

Mr. Sylvan Goldman of Oklahoma City invented the shopping cart in 1937. Mr. Goldman owned a grocery store, and every day he would see shoppers lugging around their purchases in small, hand-held baskets. He believed that wheeled carts would make shopping easier for his customers and thus attract their business.

When Goldman first put his wheeled carts at the disposal of his customers, he was both disappointed and puzzled to find that no one would use them. Not only were they in the habit of carrying a basket, many were offended at the idea that they might not be strong enough to carry their purchases.

Goldman was sure his carts would be a great success if only he could persuade people to give them a try. To this end, he did something very clever and amusing. He hired bunches of people to push carts around his market and pretend they were shopping! Seeing this, the real customers gradually began copying the phony shoppers.

Stethoscope

❏

The stethoscope was conceived in 1816 by Dr. Rene Laennec while examining a young woman with a heart condition. The young woman was very pretty, and Dr. Laennec was too shy and modest to put his ear to her chest. To solve the dilemma, Laennec fashioned a make-shift listening device from a piece of paper rolled into a tube. "I was both surprised and gratified," he wrote, "at being able to hear the beating of the heart with much greater clearness than I have ever done before." In time, Laennec constructed a wooden stethoscope, variations of which soon became a standard of the medical profession. When a book written by Laennec went into print in 1819, the publisher gave a free stethoscope with each copy sold.

Submarine

❑

The first practical submarine was basically an underwater rowboat. Built in London in 1620 by a Dutchman named Cornelius Drebbel, the outer hull of the craft consisted of greased leather over a wooden frame. Oars extending through the sides and sealed with tight-fitting leather flaps provided the means of propulsion.

Between 1620 and 1624, Drebbel, on several occasions, successfully navigated his craft up the Thames River at depths of from 12 to 15 feet. On one occasion, King James I of England went aboard the vessel and took a short ride.

Sundae

❑

In Evanston, Illinois, in 1875, a law was passed forbidding the sale of ice-cream sodas on Sunday. To get around the law, on Sundays enterprising soda jerks began serving ice cream with syrup but no soda water. The concoction became popular, and on weekdays customers began asking for "Sundays." City officials objected to naming the dish after the Sabbath, so the spelling was changed—and it has been "sundae" ever since.

Sunglasses

❑

In ancient China, judges wore smoke-tinted glasses to conceal their eye expressions. Real sunglasses—those designed to reduce solar glare—were invented in the 1930s by the Army Air Corps. The company of Bausch & Lomb was commissioned, and soon came out with green-tinted sunglasses to shield aviators' eyes. During the 1950s, sunglasses were becoming popular all over the United States and the rest of the world.

Supermarket

❑

The first self-service grocery store was the Alpha Beta Food Market, which opened its doors in 1912 in Pomona, California. Shortly thereafter, a chain of self-service groceries known as Humpty Dumpty Stores was started in California by the Bay Cities Mercantile Company.

In 1916, C. Saunders opened the Piggly Wiggly supermarket in Memphis, Tennessee. Saunders introduced a turnstile entrance and a regular checkout system. To make sure people saw all his goods, Saunders organized his store in such a way that shoppers had to go up and down every aisle before reaching the checkout counter. Within seven years, Saunders had built up a chain of 2,800 Piggly Wiggly stores throughout the United States.

Teddy Bear

❏

In 1902, President Teddy Roosevelt made a trip to the South. While there, he went on a hunting trip sponsored by his Southern hosts. Wanting their honored guest to return home with a trophy, they trapped a bear cub for him to kill. Roosevelt was disgusted. He refused to shoot the helpless little animal.

The story gained national attention. A cartoon in the *Washington Star* showed Teddy Roosevelt, rifle in hand, with his back turned on a cute, cowering baby bear.

Morris Michtom, owner of a Brooklyn toy store, was inspired by the cartoon to make a stuffed bear cub. Intending it only as a display, he placed the stuffed bear in his toy store window, and next to it placed a copy of the cartoon from the newspaper. To Michtom's surprise, he was besieged by customers eager to buy a "Teddy's bear."

Michtom was soon manufacturing Teddy bears by the thousands. The proceeds enabled him, in 1903, to form the Ideal Toy Company.

Tennis Shoes

❑

Around the turn of the century, rubber soles were being glued to canvas tops to make lightweight athletic footwear. In 1917, the U.S. Rubber Co. introduced Keds, the first popularly marketed tennis shoe, choosing the name because it suggested "kids" and rhymed with *ped*, Latin for "foot."

In the 1960s, Phil Knight, a miler at the University of Oregon, wanted tennis shoes that had more traction. One morning he took a waffle iron, put a piece of rubber in it, heated it, and produced a waffle-shaped pattern. Soon Knight and a partner went into the business of manufacturing a new type of running shoe, one with a tread and higher-traction sole. He called the shoes Nikes, after the winged Greek goddess of Victory.

Thimble

❑

To protect their thumbs when sewing heavy canvas sails, British sailors wore little leather sheaths called "thumb bells." In the seventeenth century, a London metalworker, John Lofting, saw one of these devices, duplicated it in thin steel, and changed the sewing habits of the age. In time, "thumb bell" was slurred by popular speech into "thimble."

Tinker Toys

❏

Charles Pajeau, an Evanston, Illinois, stoneworker, came up with the idea of Tinker Toys in 1913 while watching some kids playing with "pencils, sticks, and empty spools of thread." He designed his first set in his garage, and with high hopes, displayed the toy at the 1914 American Toy Fair. But nobody was interested. He tried his marketing skills again at Christmastime. He hired several midgets, dressed them in elf costumes, and had them play with "Tinker Toys" in a display window at a Chicago department store. This publicity stunt made all the difference in the world. A year later, over a million sets had been sold.

Toothbrush

❏

The year 1770 found William Addis serving time in an English jail. One morning he was cleaning his teeth in the way people had for centuries—by rubbing them with a rag. As he did, an idea came to him. He found a small piece of bone and a brush. He bored tiny holes in one end of the bone, cut some bristles down to size, wedged them through the holes, and glued them in place. When he was done, he had in hand the world's first toothbrush.

Soon after being released from prison, Addis went into the toothbrush-making business. The business was an overnight success, and Addis was soon a wealthy, honest man.

Traffic Light

❏

The world's first traffic light came into being before the automobile was in use, and traffic consisted only of pedestrians, buggies, and wagons. Installed at an intersection in London in 1868, it was a revolving lantern with red and green signals. Red meant "stop" and green meant "caution." The lantern, illuminated by gas, was turned by means of a lever at its base so that the appropriate light faced traffic. On January 2, 1869, this crude traffic light exploded, injuring the policeman who was operating it.

Half a century later, Garrett Morgan of the United States realized the need to control the flow of traffic. A gifted inventor and reportedly the first African American to own an automobile in Cleveland, Ohio, he invented the electric automatic traffic light. Though it looked more like the semaphore signals you see at train crossings today, it provided the concept on which modern four-way traffic lights are based.

Twinkies

❑

Twinkies were invented in 1931 by James Dewar. He originally called them Little Shortcake Fingers, but while on a business trip, Dewar and a friend passed a shoe factory sign that read: "Home of Twinkle Toe Shoes." Dewar had been looking for a new name for his product. His friend suggest Twinkle Fingers. Dewar shortened this to Twinkies.

Typewriter

❏

In 1872, Christopher Sholes and Charles Glidden were developing a machine for numbering book pages. Glidden asked one of those wonderfully simple questions which are often the basis of inventions: "Why cannot the paging machine be made to write letters and words, and not numbers only?" The result was the typewriter.

Though it had no shift key and printed only in capital letters, it worked. The apparently mad scattering of the alphabet on the keyboard (which remains with us to this day) was so arranged to prevent jamming. In the first typewriters, pushing the keys caused a long rod to flip up and type the letter on the paper. Sholes put those letters which most commonly occur together when we write as far apart as possible on the keyboard.

Mark Twain was the first author to ever own a typewriter. His famous book, *Huckleberry Finn*, was written using one of Sholes' machines.

Underwear

❏

People did not begin wearing underwear until the middle of the nineteenth century. Prudishly, they were referred to as "unmentionables," "indescribables," and (perhaps silliest of all) as "unwhisperables." Underwear then was always white, usually starched, often scratchy, and made chiefly of calico, flannel, or wool. Silk underwear for women was not introduced until the 1880s.

For most of its history, men's underwear was of the "long john" variety. In 1934, it was revolutionized with the introduction of the jockey brief. The Wisconsin firm of Cooper and Sons copied the design from men's swimsuits then popular in France.

Velcro

❏

In 1948, Swiss mountaineer George de Mestral was on an Alpine hike. As was often the case for him, he was annoyed by the thistles and cockleburs that clung to his pants and socks. While picking them off, he suddenly had an idea. He thought it might be possible to produce a fastener based on the burrs.

Mestral took his idea to a weaver in France. Working by hand, the weaver was able to produce two strips of fabric, one with tiny hooks, the other with a surface of even smaller loops to be snagged by the hooks. Mestral christened the sample "Locking Tape." Later, for a trademark name, he came up with Velcro. "Vel," the first syllable of velvet, was combined with "cro," the first syllable of crochet.

Vending Machines

❑

The first vending machines appeared in England in 1615. Putting a penny in the slot released a pipeful of tobacco to the customer. Since then, vending machines have been dispensing hundreds of other things. The range of goods has included eggs, biscuits, perfume, handkerchiefs, towels, cough drops, and accident insurance. In the mid-1890s, the citizens of Corinne, Utah, were able to obtain divorce papers from a machine, and in Berlin, Germany, in 1924 a vending machine was in operation that would dispense a valid college degree.

Perhaps the strangest—and silliest—use of a vending machine was one installed in 1960 in Macy's department store. Throngs of people came to see it, but almost no one used it. They just laughed at the thing. After all, who wants to buy men's underwear from a vending machine?

Wedding Ring

❏

Have you ever wondered why married people wear wedding rings, or why the ring is always worn on the third finger of the left hand? The custom comes to us from ancient Egypt. To the Egyptians, the ring was a symbol of eternal love, for all circles are without beginning or end. If the ring broke, it meant bad luck. If it were taken off, love would escape from the heart. The Egyptians believed that a long vein ran from the third finger of the left hand to the heart. That is how the third finger, left hand, became the ring finger.

Wheelchair

❑

In 1651, J. Haustach of Germany put his mind to helping a friend who had lost the use of his legs, and in so doing became the inventor of the wheelchair. Haustach took an adult-sized tricycle and modified it by making a larger, more comfortable seat. Then he added a big front wheel that had an internally toothed gearbox. Sitting in the tricycle-like contraption, his friend worked two hand cranks. Pulling them back and forth made the front wheel turn and the chair go forward.

Wigs

❏

To cover baldness, or for other reasons, wigs have been used since the beginning of recorded time. Egyptian men and women wore pads of false hair as protection against the burning rays of the sun. In Rome during the first century B.C., women—but not men—began wearing hairpieces. During the sixteenth century in Europe, it suddenly became fashionable for both men and women to wear wigs. So great was the demand for wigs at this time that it was not uncommon for children to be lured away and robbed of their hair!

By the eighteenth century, wigs had become outrageous. Women's wigs were sometimes four feet high! These ridiculous headdresses were dusted with flour and decorated with phony fruit, stuffed birds, and even model ships!

Windshield Wipers

❑

The first windshield wipers had to be operated by hand! Invented in England in 1920, either the driver or a passenger had to work a crank to make the wipers go back and forth.

Automatic windshield wipers were invented the following year. Called "Folberths," after their inventor, W. M. Folberth, they were powered by an "air engine," a device connected by a tube to the inlet pipe of the car's motor. Electric windshield wipers were invented in the United States in 1923.

Wire Coat Hanger

❏

One morning in 1903, Albert J. Parkhouse arrived as usual at his workplace, the Timberlake Wire and Novelty Company in Jackson, Michigan, which specialized in making lampshade frames and other wire items. When he went to hang his hat and coat on the hooks provided for the workers, Parkhouse found all were in use. Annoyed—and inspired—Parkhouse picked up a piece of wire, bent it into two large oblong hoops opposite each other, and twisted both ends at the center into a hook. Then he hung up his coat and went to work.

The company apparently thought it was a good idea, because they took out a patent on it. In those days, companies were allowed to take out patents on any of their employees' inventions. The company made a fortune; Parkhouse never got a penny.

Wristwatch

❑

Wristwatches were invented in Switzerland in 1790. Originally, they were called "bracelet watches," and were only for women. The first wristwatches for men did not come into being until almost a hundred years later, and men who wore them were considered sissies. Not until shortly after the end of World War I (around 1920) did wristwatches for men gain acceptance.

Yo-yo

❏

The yo-yo was originally a hunting weapon used by Filipinos. It consisted of a large disk of wood or stone around which twine had been wrapped. The weapon was hurled, and the twine snared an animal by the legs. "Yo-yo" is a word from Tagalog, the native language of the Filipino people. In English it means "come-come."

In the 1920s, an enterprising American named Donald Duncan, on a visit to the Philippines, happened to see a hunter using a yo-yo. Scaling down the size of the device, he transformed it into a toy, and was soon selling them in the United States by the thousands.

Zipper

❑

In 1893, Whitcomb Judson decided to invent a replacement for shoelaces. Eventually he came up with what he called "The Clasp Locker and Unlocker," a device consisting of two metal chains that could be joined together by pulling a slider between them. This first crude zipper did not work very well on shoes—or on anything. The U.S. Postal Service bought twenty mailbags with Clasp Lockers, but they jammed so often that the bags were discarded.

By 1910, Judson had designed a new, improved fastener. Called the C-Cuity, it sold for 35 cents. The C-Cuity was not for use on footwear, but on men's pants and women's skirts.

One day a businessman happened to be visiting Judson's factory. Judson demonstrated how the fastener worked. "My, that's a zipper!" exclaimed the businessman, using a popular slang term of the time.

From that day on, Judson's little invention had a new name. And that, as you know, is a zipper!